You must read this book in chronological order from beginning to end to get the benefits from it. Otherwise, you will not understand the content. It is written in a way that requires it.

Introduction

Do you want to truly know where you come from? Well, you have chosen the right book. Not only will this book answer that question, but it also gives you the understanding of who and what you are. As a result, you will understand yourself, others, and things around you significantly more than you do, presently as well as comprehend more than others who have not read this book.

This is not like any other information that you have heard. You will learn this newfound secret that I have discovered or that has come through me.

This book gives you a rational or logical understanding based on something that you can see. However, this is not a science book. Nor is it a book based on theology or religion. It is based on the foundation of "what is". Although at times, I will use examples from basic science to explain points that you need to know to see the big picture. This is not to shed a negative light on science and religion. Science and theology are good, and you don't have to give them up to see what I am going to tell you. In fact, in some ways they may help you to understand what I am going to tell you.

You have probably chosen this book because you have not found a satisfying answer from any branch of science or any topic of religion. This book goes beyond the scope of any branch of science like genealogy and points out things that theology does not.

Since you refer to yourself as a human being or homo sapiens, it shows that you have accepted the answer that science has given you to some degree. It is evident to me that scientists do not know where you come from by reason of these words used to define you. In this case, they don't know "what you are" either. How can you know what you are if you don't know where you come from? Consequently, you are left not knowing where you come from, nor what you are. If this is true, you are left with a deeper dilemma which leaves an additional more intimate question. Accordingly, if you don't know where you come from nor what you are, you don't know who you are. So, you should be asking, "who am I?" On these grounds, you must rule scientists out when you are attempting to find the answer to this question. After reading this book, you will understand the answer to all these questions: where did I come from, what am I, and who am I?

Like you, I have read many books to try to find the answer. Books of science is one of the options I took to try to discover a satisfying answer to the question: Where did I come from? In fact, my major in college is pre-medicine and I worked as a physician's substitute for a few years. So, I have a science background. Therefore, I have read my share of science books in school as well as for my personal learning. Basically, you end up with the big bang theory, evolution, and the sperm and egg or reproduction. Firstly, while the big bang theory is a good theory, it is a bit abstract. If you really think about it, this whole concept is like the Holy Bible when it speaks about God speaking worlds into existence. It is powerful but not quite what I was looking for. Secondly, you have evolution. I'm convinced that the universe, God, or whatever you feel comfortable with, has power or ability to create humans and other creations without having to go through this channel or theory of evolution to do it. Besides, the big bang theory and evolution leaves new created life in a state of vulnerability. If you are newly formed from a sperm and egg, you wouldn't last long after being created because new living things really couldn't take care of themselves to survive. Thirdly, reproduction leads you to genetics or genealogy which has a limited scope. In other words, they can tell who your ancestors are, but it does not tell you how they got here in the first place. It is limited to telling you which line of humans you come from. Again, this is an incomplete and unsatisfying answer to the question.

Then, you have theology and/or spiritual resources. I was practically raised on theology. My father was a preacher and one of my uncles was and still is a pastor. I have been going to church since I was born. As a result, I know a lot about the subject of religion and I have read a lot of spiritual books to try to find the answer. There are many different types of religions. I have studied a lot of them. When you compile them, they all tell you the same thing in a different way. They tell you that God or the universe created you. Consequently, that does not really answer the question. You are still left wondering how this happened and even where this happened. It really doesn't answer the question because the answer is abstract. It is not something that you can really see or visualize. Unlike the big bang theory and evolution, theology and spiritual teachings have you created as a full-grown living thing and are able to survive since God is telling you what to do. However, when you look around and think about life, you can see that there are things missing from this scenario. If you believe in God or the universe, you can see that there is more to it than what is revealed. You can easily see that God, or the universe is cleverer than that. You are not getting all the answers. It is incomplete and abstract. For this reason, you are still left wondering. Theology and spiritual teachings tell you that God or the universe created you and you are made in his image. Unfortunately, you don't have any images of God to be satisfied with this answer. If you try to base it on the things you see around your milieu, you become more perplexed. On these grounds, you must rule theology and spiritual teachings out when you are attempting to find the answer to this question.

In summary, science has the big bang theory, evolution, and genetics that leave you wondering which came first, the chicken or the egg while religion has many types that leave you with incomplete abstract ideas. Since you are not abstract, this makes it hard to grasp and is dissatisfying. Neither avenue really reveals where you come from.

Since this is true, you are like an adopted child. That's right because adopted children do not know where or who they come from. This is what I will reveal to you. Since you are like an adopted child, you don't understand who you are. This can be a challenge mentally or merely frustrating. As an adopted child, you feel as though you have a void in your life. If this was not true, you would not be searching to find out where you come from. This may affect you without you realizing the effects. You have been looking for answers and have been unsuccessful. The point is, something is missing and needs to be found. Innately, it is something that we all long to know.

If finding out the answer to this question is not important, you wouldn't have adopted children and children who don't know who one of their parents are searching all over the world to find them. You wouldn't have archeologists digging up tons of dirt searching for it. You wouldn't have an entire branch of science devoted to it.

Finding out will enlighten you. It will be transcendental for you. Like an adopted child, you will see the universe and life in a different way once you find out where you come from. In the event you find out, you will learn who and what you are. This answer goes beyond science and religion and is not political. Reading this book will fill the void in your life that you may not even realize that you have. Likewise, it will give you closure. It will help you understand why you do certain things and help you to see things more clearly as they really are. You will have started a new journey of finding out "what is".

Gaining knowledge of where you come from is a learning process. In other words, there are a few things that you need to understand to really comprehend and accept where you come from. Like an adopted child, finding out may be a challenging, traumatic or frustrating experience. By contrast, it may be the most delightful thing you ever experience by knowing. Therefore, it is important for you to read this book in chronological order from beginning to end and not skip around the chapters. If you skip around the chapters you won't understand why I'm saying certain things and you will not get the benefits from this book. It is like you're trying to learn multiplication without first knowing what numbers are. It simply won't make since and will not resonate. I will explain things that you need to know while revealing where you come from. As I stated before, this is something that you have not heard before. It is new to you. So, take your time and let it unravel. You may need to wrap your mind around it, at first. Give yourself a fair chance to learn to information the way it was intended. Rewardingly, you will learn the answer to the question and more, if you follow these instructions. As a result, you will learn the fundamentals which will open your mind to understand where you come from when I reveal it to you.

Learning where you come from is to learn what and who you are. Since this is the case, you will have a new understanding of yourself after reading this book.

Stages

As you know, your life comes in phases or stages. Firstly, you are combined as the sperm and egg. Secondly, you are a fetus. Thirdly, you are an infant. Fourthly, you are a child. Fifthly, you are a teenager. Sixthly, you are a young adult. Seventhly, you are middle aged. Eighthly, you are a senior. This list of phases in life may vary depending on whose list you review. Each phase has a physical as well as a mental stage. It is as important to know where you are collectively as well as individually. Knowing what stage you are in collectively is to know what stage you are in all together. Meaning, all people combined on this planet are in a stage or phase in life. Although it may seem different, the stage in life collectively may be related to the stage in life individually. For the scope of this book, it is important to understand this from a mental or intellectual capacity collectively. You will see that the impact is greater.

Collectively, you have figured out things like physics, engineering, construction, mathematics, music, medicine, mechanics, robotics, flight, agriculture, physiology, surgery as well as many others just to name a few impressive things that you have done. Keep in mind that it took centuries to accomplish this. Collectively, this would be from the mental or intellectual capacity of an adult. Although, I am not saying that this is the stage you are in collectively.

On the other hand, you do not know where you come from. Collectively, you have been unable to answer this question. This is an important question that starts being asked at the stage of a toddler or child. A child may ask their parents, where they come from.

In summary, you are advanced intellectually, but you don't know where you come from. You can figure things out without anyone really telling you how to do it, but you don't know who or what you are. This makes you a prodigy, collectively. You have a high IQ, but you don't know much about life. You are a like a child genius. Having a high IQ does not make you have the mind of an adult. An adult would have knowledge and understanding of such basic things like what you are. If you don't know where you come from, you don't know what you are. Since this is the case, you are at the stage of a child. This is true because collectively this information and much more is not known at the time I am writing this book. Also, it takes time to move through stages. Avoid not accepting the stage or phase you are in to learn where you come from.

It is important to understand this concept to fully accept my guidance through this learning process. Simply open your mind up like a child who is eager to learn new things. Allow me to take you by the hand step by step.

Do You Like Categories?

For centuries, we have named and categorized things. According to the Holy Bible, we have done it since the beginning of the creation of man. The book states that God gave man dominion over all living things and gave him the authority to name all things. Furthermore, scientists name and categorize things quite readily. Often, we do this even when we don't really know or completely understand what we are naming or categorizing. For example, like when scientists make a new discovery they proceed to name, categorize, and define it. This is helpful when you are communicating to others. However, it is challenging when you are attempting to learn new things about something that has already been named, categorized, and defined. The reason for this complication is because with words meanings come. When you give something a meaning that you don't fully understand, it may require renaming or even deleting the name or category from your mind. Otherwise, it leads you to learn it incorrectly. You have been forced to see and believe it to be a certain way. At the same time, you believe what you have learned which is what can complicate the process of learning it the correct way. It may cause you to be challenged with seeing it in a different way after learning it in the incorrect way, at first. It's like a child who has learned that stealing is good. It would be a challenge to try to teach them that it is not good to steal if they still held that belief system into adulthood. It may be less of a challenge if they are still a child and have not stolen much. Nonetheless, it would be a test to teach them the opposite.

This is the contest that I am faced with while I am trying to explain this information to you. It is important that you understand this as I will be explaining things to you about certain subjects that you have learned in a different way. I may change the name so that you are inclined to look at it in a different way. I will let you know in advance when I will be doing this. Just remember that it is okay to look at things in different ways or unlearn the way you presently see things. If you don't understand this, it will probably stifle the true vision in your mind eye.

From what I have been told, this is the reason that some native Americans don't name their children when they are born. They wait until they see their personality and learn what they are about, first. Then, they proceed to name them after the child has defined themselves so to speak or have given themselves meaning. This is because they know with names come meanings. They want it to resonate with the nature of the child or with their being. This circles back to what I was saying about naming and categorizing something without really knowing or understanding what it is.

In summary, giving something a name or definition against its' true nature creates a veil to the truth. This can prevent you from finding the real answer which will prevent you from finding additional hidden truths that you did not know would be their when you uncovered the truth. Avoid naming and categorizing things in order to learn their true purpose.

Does One Size Fit All?

It will be important to understand the relevance of size to comprehend where you come from. This topic may seem obvious at first, but it needs to be pointed out. As you know, things that are large and things that are small. The things that I am referring to are living things. There are living things around you that are large like whales while things that are small like gnats. Then, you have things that are referred to as microscopic organisms. Finally, you have things that seem very large like what you call planets. By the way, it is critical to understand that these categories or names like microscopic organisms and planets can impair the vision in your minds eye. Things simply are the size they are. If you look at them any other way you will miss the big picture of what they are as well as their nature. If you see yourself the same, it clears your vision. If something is considered small or microscopic and you want to understand it, see yourself as small or microscopic. The same is true if the thing is large. It is easy to see things on the same plane as you because you already see things from that perspective. It's similar if you need to understand the motive of someone who is poor while you are rich. You must visualize yourself the same and you will begin to understand things from their perspective. The criminal justice department understands this principle and applies it when they are attempting to find the motive of a suspect. According to correctionsone.com, there is an identified beginning to criminal behavior, and it starts with biology and genetics. This shows that you really have to put yourself at the level of the thing you are studying.

The timing compared to another living thing is relative to its size. However, the timing of a living thing is relatively the same when compared to itself or others like it. In other words, the smaller something is in comparison to another thing exponentially, the faster it seems to move and the shorter it's life span. By contrast, the larger a living thing is in comparison to another living thing exponentially, the slower it appears to be and the longer it's life span. For example, a gnat moves much faster than a whale by comparison and a gnat has a much shorter life span than a whale as well as when you measure their life agedness. This is consistent with all living things in nature: the larger it is to you the slower it seems and the longer it's life span while the smaller it is in comparison to you, the faster it moves and the shorter it's life span. Let's separate this. The larger it is to you the slower it seems while the smaller it is in comparison to you, the faster it moves. This is the Law of Fast and Slow Motion that I discovered. Then, the larger it is in comparison to you the longer it's life span while the smaller it is in comparison to you, the shorter it's life span. This is the Law of the Span of Life that I discovered. When I speak of these laws, I am talking about a relatively large difference in size like a gnat or in comparison to you or a planet in comparison to you.

At the same time, the small living things movements and lifespan are average or approximately the same as others in their species when viewed from their own perspective. To you, the life span of the gnat is around 14 days which seems short, but to them it is a full life span that is equal to yours. Respectfully, your lifespan appears to last for centuries from the gnat's perspective. Be as it may, your lifespan seems average or even short to you. Amazingly, this is where fast and slow motion originates. All the same, with this dynamic in place, there is no such thing as time. This may seem confusing, but there is a reason for this. I will go further into this later. It will be clear when I apply it as needed.

To you, the gnat seems small and the whale appears large. From their perspective, they are sized appropriately which is average. Likewise, when you compare other people, you seem to be sized appropriately. The smaller a living thing is in comparison to another living thing, the more different or foreign they tend to look from each other. Since they look foreign, this means they tend to look unrelated. For example, a microbe looks foreign to a human being. This is also true with the situation reversed. Then, a human being looks foreign to a microbe. Therefore, a human being and a microbe looks unrelated. This is something that is consistent in nature with all living things. This the Law of Appearance According to Size that I discovered. I will explain the relevance of this later.

In summary, it is important to understand the relevance of size. You must visualize yourself the same and you will begin to understand things from their perspective. The timing of one living thing compared to another living the is relative to its size. Living things with drastic size differences tend to look unrelated.

Can You Duplicate it?

This is another topic that may seem obvious, but it needs to be pointed out to make a point later in this book. You will be learning some fundamentals of replication. Reproduction is what keeps living things in existence. Ultimately, if what we call a "species" does not live forever individually, they must procreate to survive. A living thing is in continuation because it has replicated at least once to continue to remain for generations. Another way to look at it is if there are many of the same species that have ancestors or descendants, they have reproduced. If you see younger smaller ones in comparison to older larger ones, they have replicated.

All living things do not procreate the same way. Trees and other similar living things recreate by dropping seeds into the soil. Some other creatures lay eggs outside their bodies and have them fertilized. We mammals, as well as others have eggs inside our bodies and have them fertilized. All three methods provide the same results. They all produce offspring.

All living things that exist were not reproduced. Instead, they were what I call conjunction created. In this case, some things are clones and some things are created along with the reproduction of another living thing. This may seem confusing at the first glance, but I will explain it to you. For example, there are micro-organisms that live on and within your body. I will refer to microorganisms as micro living things. These micro living things were conjunction created when you were reproduced. They come with you and belong to you exclusively. This is the reason why your body rejects things that were not originally a part of you like transplants or donor organs. It is also important to understand that their make-up is a result of your parent's genetics. Your parents are their parents too. Yes, I know that it may take a minute for it to wrap around your head, but you read that correctly. It may sound strange, however the micro living things that live in and on your body are related to you. They are the offspring of your parents too. This is the reason that close relatives are more likely to be a match for things like bone marrow, transplants, transfusions, etc. They come from the same line of ancestors. Nature is complex.

When most living things reproduce, their young offspring are usually nearby. Notice that I said most living things have their offspring nearby. There are outliers with this fact. For instance, this does not apply to turtles. When they lay their eggs on the beach shoreline, they leave them to fend for themselves after they hatch. There are a few other species that act in a similar manner. However, this is not the norm for the offspring of most living things.

Earlier, I mentioned that living things with drastic size differences tend to look unrelated. I don't mean in comparison to you and a gnat because there are still familiar similarities. I mean like the micro living things that live on and in you in comparison to you. That is because things are created to fit their environment as opposed to being created to look similar.

In summary, it is important to understand that everything reproduces. Different living things reproduce differently. Some things are created along with the reproduction process or conjunction created. Most living things keep their young offspring nearby. To learn where you come from, don't believe that the offspring is not near their parent. You will understand the relevance of this later.

Can You Learn It?

It is important to learn about where you are to understand where you come from. It is necessary to discuss a little about this planet that you call earth and this solar system. It is critical to not name or categorize them like that in your minds eye. I will use different words to communicate with you to help diminish your clouded vision of these things.

It's crucial to look at this planet as well as the other planets in this solar system simply as "living things" instead of planets. I will refer to them as "spherical living things" as I refer the earth as "this or our spherical living thing" and this solar system as our area. I will also be talking about the sun which I will refer to as "fiery spherical living thing". The moon will be included as well, and I will refer to it as the "illuminating spherical living thing".

Firstly, we will discuss the conditions of these spherical living things. These spherical living things are naturally without light. Without the fiery spherical living thing, they would be in darkness. Darkness is their natural state. Since we have ruled out light, then there would be no heat. Without light there is no heat. If this is true, their natural state is also to be in an extreme state of cold. I am speaking of hundreds of degrees below zero. According to Wikipedia, the coldest temperature ever recorded is -89.2 degrees Celsius (-128.6 degrees Fahrenheit, 184 degrees K) at the Soviet Vostok Station in Antarctica on July 21, 1983 by ground measurements. This demonstrates that it is very cold even with the fiery spherical living thing present. This makes it easy to see that it will be much colder without the fiery spherical living thing existent. It would be hundreds of degrees colder than it is, now. In this case, this extreme cold brings solidification. This brings these spherical living things to be in a solid state or frozen.

By contrast, this makes it obvious that light is the natural state of the fiery spherical living thing. According to Cool Cosmos as well as other science resources, the temperature on the surface is 10,000 Fahrenheit (5,600 degrees Celsius) while the temperature at the core is 27,000,000 Fahrenheit (15,000,000 Celsius). In this case, the sun brings extreme heat. This hot nature could not possibly produce cold. This means that fire does not belong here on this spherical living thing as fire produces light and heat which is not natural to it. Therefore, fire does not belong or come from our spherical living thing.

Since spherical living things are normally in a solid or in a frozen state, weather is not ordinarily here. If this is true, it rules out seasons. Seasons are not natural to spherical living things.

Secondly, what you call orbiting is not a natural constant movement for spherical living things. It is crucial to not name or categorize this movement as an orbit to learn where you come from. For the purpose of this book, I will refer to this movement as circling. Their movement depends on what is happening at the time. Just like your movements, it depends on the time and place or circumstance to determine the occurrence.

According to Wikipedia, our spherical living thing is 4.543 billion years old and the sun formed 4.603 billion years ago. The illuminating spherical living thing is 4.53 years old. In this case, the fiery spherical living thing is older than our spherical living thing while the earth is older than the illuminating spherical living thing. I will expound on this topic later but keep in mind that these are thousands of years in difference between them.

According to space.com, the radius of our spherical living thing is 3,959 miles which is 6356 kilometers. According to solarsystem.nasa.gov, the radius of the fiery spherical living thing is 432,168.6 miles which is 695,508 kilometers. According to Wikipedia, the illuminating spherical living things radius is 1,079 miles. As you can see, the fiery spherical living thing is much larger than our spherical living thing while our spherical living thing is larger than the illuminating spherical living thing. I will go into further detail about this later.

In summary, our spherical living thing's natural state is frozen while the fiery spherical living thing natural state is extremely hot. Also, our spherical living thing is small in comparison to the fiery spherical living thing while the illuminating living thing is small in comparison to our spherical living thing.

Why Is It Paradoxical?

I mentioned earlier that everything reproduces. For this to work, you need a male and a female. Well what I discovered is, the fiery spherical living thing is a male while the other spherical living things are females that are in our area. Males and females mate or have courtship that includes intimacy. The intimate part of this relationship may include different things like a mating call and certain movements, almost like a dance. This is evident with the fiery spherical living thing and the spherical living things. I cautioned you earlier to not refer to our spherical living thing to be orbiting the fiery living thing. Instead, it is circling the fiery spherical living thing. It is more like a dance. This is how they court, mate, or be intimate. Earlier, I mentioned that the timing of larger things appears to be moving slower to you and they live longer by comparison to living things that are smaller. To you, this circling while spinning movement is a very slow motion, but it is fast from their perspective.

The fiery spherical living thing is hot while our spherical living things natural state is cold. As you can see they are opposites. When their bodies unite in this special way of circling closely, their opposite natural states sort of combines. They don't have to be touching like you do when you mate. This combining is evident with this or our spherical living thing being able to produce fire and having a hot core. The fact is, you can start a fire in humid conditions or even in the winter. Remember, without the fiery living thing present, the spherical living thing is frozen solid. Therefore, fire and a hot core are not possible.

The process of reproduction often creates offspring. As you know, the illuminating spherical living thing has ability to light up or illuminate. I know what scientists say about the light from it, but this is irrelevant. They say that it is in a way caused by the fiery spherical living thing casting light and it illuminates. We are interested in the fact that it has ability to illuminate or do something with light no matter how it can do it. This is important because the fiery spherical living thing casts light while our spherical living thing does not. Since the illuminating spherical living thing has ability to do something with light means it falls between the fiery spherical living thing which produces light and our spherical living thing which does not produce light. The reason the illuminating spherical living thing has ability to do something with light is because it is a combination of the fiery spherical living and our spherical living thing. In other words, the moon is the offspring of the fiery spherical living thing and our spherical living thing. If this is true, then the fiery spherical living thing impregnated our spherical living thing. Remember, it was established earlier that the illuminating spherical living thing is younger and smaller than our spherical living thing. Also, the illuminating spherical living thing is much smaller than our spherical living thing. The illuminating spherical living thing is young, so it has more growing to do. The illuminating spherical thing is not circling our spherical living thing to give you light at night. They are staying near their parents. Mainly, the mother. Kids are more inclined to be around the mother a lot more. Especially, since it is a mixed offspring. What I mean is, even though it is mixed, it is still quite different in temperature to the father. Consequently, they can only tolerate a certain amount of heat from the fiery spherical living thing. Therefore, they cannot get too close. The illuminating spherical living thing probably can get a lot closer than our spherical living thing.

When you look at our spherical living thing from afar to the point where you see it as a sphere form a satellite, you can not see life or people. In this case, by comparison to our spherical living thing, you are microscopic. Since this is true, you are a micro living thing. I mentioned earlier, that some things are conjunction created and this applies to you. If something is impregnated, it carries life. The fiery spherical living thing caused our spherical living thing to carry life. When he did this, the illuminating spherical living thing was created, and you were conjunction created. If this is true, then you are related to the illuminating spherical living thing. All the while, the fiery spherical living thing and our spherical living thing are the parents of the illuminating spherical living thing. Since this is the case, the parents of the illuminating living thing are your parents too. Yes, you read that correctly. That finally answers the question. That is where you come from. I know what you are probably thinking or what you will think of later. You believe or will think that you would have been conjunction created on our illuminating spherical living thing. This is not the case. They do have conjunction created life, but it is not you, of course. Also, you may think that you were conjunction created when our female conjunction parent was born. However, this is not possible. I stated earlier that her natural state is frozen hundreds of degrees below zero. This is true because she is not mixed or biracial or created from a fiery spherical living thing. You will understand this better shortly when I explain your likenesses. Nevertheless, since her natural state is hundreds of degrees below zero, you would not survive here under that condition. You would freeze solid.

I mentioned earlier that living things with drastic size differences tend to look unrelated. You look unrelated by appearance, but your make-up and characteristics are not far apart as I will explain shortly. This is because you are designed to fit the environment of where you are or live, not to where your conjunction creators are or where they reside. Therefore, you look different from them on the outside. For example, you look different from our spherical living thing as micro living things that live on and in you look different from you. Also, earlier I said that some living things recreate differently. If this is the case, they also conjunction create differently as well. Since this is true, that is the reason that you were conjunction created on our spherical living thing and not on the illuminating spherical living thing.

Also, with timing, your spherical and fiery spherical living things timing or life span appears longer to you so that you can live here for generations.

It is important to explore the fact that we live in a world of opposites to understand where you come from. One of the reasons you have paradoxes is due to the opposing nature of your conjunction creators or conjunction parents. Our conjunction parents couldn't be more paradoxical. This is the most-odd couple that you will find. As I said before, the male is extremely large in comparison to the female. This is the reason that males here tend to be larger than females, normally. It is in your genes.

Earlier, I mentioned that your male conjunction parent is older than your female conjunction parent. This is the reason females here normally prefer males that are a little older and men like women who are younger. This is a common likeness that you have in common to your conjunction creators.

I said before, our female conjunction creator is cold, and our male conjunction creator is hot which makes them total opposites. This is the reason opposites here attract. Also, this is the reason things here depend so much on our male conjunction parent. In other words, you need him to survive. His rays cause what we know as photosynthesis and the production of vitamin D when your bare skin is exposed to it, for example. There are many other benefits from the rays of our male conjunction creator that we depend on. This doesn't occur simply because it is a fixed solar system like scientist believe. This happens because your survival is dependent upon him because you're the living thing that was conjunction created. All conjunction creations survival depends on the creators of the conjunction creations or living thing that was created. This is true for all living things created. This is the Law of Conjunction Creations Survival. That is the real reason that you need your male conjunction parent.

Our male conjunction parent caused our female conjunction parent to carry or create our life while our life is dependent upon him as well as her. We live off our female conjunction creator as everything is here all that you need. Since this is true, the male conjunction parent is the provider and the female conjunction parent is the nurturer. The same thing is true for the common likeness carried down to males and females here. Males are mainly the ones who create and use to be the only providers while females are mainly the nurturers by nature like breast feeding and caretaking.

Since one conjunction creator is hot and the other is cold, this is the reason hot therapies and cold therapies are so beneficial to you as well as different types of light therapy or treatments.

Since your conjunction creators are opposite in temperature, this creates weather and conditions that is beneficial to living things here. The combination of hot and cold creates rain which is good for crops, our bodies, etc. which are crucial to your survival. It also creates a windy environment which blows pollen and other things in the air that is critical to your and other living things survival as well. Without this world of opposites, there would be no seasons. This causes it to be hot and cold and everything in between which is caused by the combination of our conjunction parents.

This world of "opposite effects" is caused by their union. These effects are carried down to everything that they create which is carried down to everything that their creations create and so on. This applies to every living thing that was ever created. This is the Law of Infinite Occurrences.

Since our male conjunction parent uses heat rays to benefit living things here with benefits like photosynthesis means that he can manipulate things with temperature. This is a fiery quality. If this true, cooking things is a quality that is passed down by our male conjunction creator. If you like to cook and eat heated food and beverages like hot cereal or any heated food, hot coffee, hot tea, hot cocoa, heated juices, etc., then you possess a likeness of your male conjunction parent. The opposite is true if you prefer cold temperature foods or beverages like cold sandwiches or finger foods, cold cereal, raw food, ice cream, yogurt, popsicles, cold coffee, cold tea, cold juices, cold water, etc. Therefore, your portrait a likeness of our female conjunction parent. This means that many animals have this likeness while plant life depicts our male conjunction creator since they benefit from photosynthesis. However, plant life that thrives in the summer is more like your male conjunction creator and the ones that do better in the winter is more like your female conjunction parent. This is not to say that livings things don't fall in the middle and have some qualities of both.

These opposites of hot and cold are the reason that you have some living things here that are warm blooded which is what you are and some living things that are cold blooded.

It is obvious that your male conjunction parent can withstand extremely hot temperatures since they can sustain them without damage to himself. If this is true, then the more melanin that you have in your skin means you have more of a likeness of him because you can tolerate heat rays with less damage than someone with a small amount. In this case, if you have a small amount of melanin means you are more like your female conjunction parent because you don't tolerate the fiery conjunction parent or his rays as well without damage. Normally, this means that you handle cold better and have alikeness of her, the non-fiery conjunction parent.

It is easy to see that your male conjunction parent's natural state is fiery hot, and your female conjunction parent's is frozen or extremely cold in her natural state. Fire has ability to grow tremendously, spread rapidly and consume things in his path and this is considered aggressive, vicious, or even violent, glaring, hot, constantly moving, and bright yet can be controlled like a flame on a candle. By comparison, ice is still, reserved, calm, cold and unmoved. However, when things get hot, she heats up. Since you are like what or who created you, then this is the reason that you tend to be more aggressive and sometimes violent but able to be tamed if you are a male here. By comparison, you tend to be relaxed, chilled, reserved, and sometimes considered cold and stubborn if you are a female. Your female conjunction parent is attracted to this fiery type of spherical living thing. In the same way, many females who are considered good girls here are attracted to males with the same likeness. Such as, bad boys, rough necks, edgy guys, etc. It is being passed down from your male conjunction parent. Your male conjunctive parent is attracted to this composed type of spherical living thing. This is the reason that many males here like girls who seem tight or reserved in the day but can loosen up at night. In other words, they heat up.

It is evident that your conjunction parents are opposites. Opposites are at different ends so to speak. Since opposites are extreme, it affects you mentally the same way according to the Law of Infinite Occurrences. This is the reason many people have mood swings. Many are affected in an even more extreme way and are diagnosed as bipolar.

In many ways, your conjunction parents have the best of both worlds. One gives fire that incinerates things in his path which is one reason why you are designed to eat, move, become excited, have heat in your body and many other things with the benefits that come with it while the other gives cold, darkness, calamity, stillness, and many other things with the benefits that come with it. You can say that they complete each other. They benefit each other. This is where prosperity originates. In the same way, you look for the same thing. In other words, one of the reasons you look for a significant other is to complete you or provide those things that you don't have is because of them. This is the reason, you call them your other half.

By contrast, the reverse is also true. Since they are opposites, in a way that one can affect the other in opposing way. It can even affect them physically. Meaning, your fiery conjunction parent can overheat your female conjunction parent while your female conjunction parent damage your fiery conjunction parent with extreme cold. In this case, they one is averse to the other. This is where your senses originate in your anatomy. In the same way, this is one cause that your health may be affected in many ways. For example, this is the reason that you get allergies and other hypersensitivities.

The union of their natural states has created a paradox or world of opposites. This has created a force that affects our terrene in many ways, not just mentally and physically. This is one reason that you have antagonists, antagonistic behavior, things can go forward and backward or be reversed, and contradictions as well as reflections which I will explain in another book called *What Gave Me Life?* to name a few.

Your fiery conjunction parent must be at a distance from your illuminating conjunction sibling. By comparison, your female conjunction parent must be even further. In other words, the father is too hot for it to be circling around him as well as other reason like nurturing that comes from the mother, therefore it circles the mother. Nonetheless, there is distance. As you know things or likenesses are passed down. Since this is true, this is the reason that fathers seem to be distant from their children here. Especially, in earlier times when only the father worked while the mother stayed home to take care of the children. Although I say they must be at a distance, it really isn't quite that far to them. It seems far to you because you are micro sized by comparison. Remember, the core of your female conjunction parent is hot, therefore she can tolerate a good amount of heat. It's just that life upon her wouldn't survive if she gets too close. Meaning, you wouldn't survive as well as most of the other living things. Yes, I did say most because there are things here that would survive such as certain micro living things. Some can survive in extreme cold as well as extreme heat including in fire.

It is evident that your fiery conjunction parent is mating with other female spherical living things. This is the reason that it is common for males here to desire more than one companion. There are other reasons that go beyond the scope of this book.

Your fiery spherical conjunction creator emits light and our spherical conjunction creator's natural state is darkness. It is obvious that these opposites bring you night and day. This is the reason that some living things are active by day while others are nocturnal. We have what is called a circadian rhythm that contributes to this which has been observed in all living things here. Circadian rhythm helps to promote sleep and wake. This occurs with you because you are biracial or have the make-up of your conjunction parents, not simply because there is light and darkness.

Since your conjunction parents are opposites by nature, this means you are mixed or what you refer to as biracial. Since this is true, this means that everyone else here is as well.

Ultimately, you come from a line of your male and female conjunction parents. Amazingly, this can be traced back by billions and billions of years.

Everything has this make-up of our area. Remember, earlier I said that I would refer to solar system as our area. In other words, the spherical living things are mating with fiery or non-fiery spherical living things and circling them. There are other areas that are like our area. These other areas are extremely large in comparison to our area and others are extremely small by comparison. Actually, our area is similar to what scientists call atoms. What they call atoms is really like our area. In other words, these atoms are merely a small or micro version of our area. This is what makes up what scientists call our galaxy. Your body has a likeness to this. This means that many of these other areas make-up is much larger, and many are much smaller than you. Since this is true, you cannot tell if you are inside or outside of another living thing. That is correct because some things live on you and in you. This is a fact for all living things. You cannot tell if you are inside because you are too small in comparison to what you are inside of. The same thing is true for living things that live in and on you. They cannot tell either. Besides, you look so different from them that it is not easy to recognize.

If this is true, there is no such thing as inside or outside. Think about it. You are living in our area which is like the atom and scientists have told you that atoms consist of or is your make-up. In other words, the atom and our area are relatively the same. Since our area is like the atom, then it is the make-up of another living thing. You are simply too small to tell what it is.

These atoms that consist of your make-up have similar living things that reside there on and within them. Since this is true, you are in or on a similar make-up of something as well just like them and so on. This goes on in each direction to infinity. In other words, living things get larger to infinity and smaller to infinity just like a number line in algebra. When a living thing procreates, another living thing is created while other living things are conjunction created on and in the thing that is procreated. Then, the conjunction creation has living things that is created on and within it and so on. Again, this goes on to infinity. In other words, micro living things that live on and in you have micro living things that live on and in them and so on. This is the Law of Creation with Conjunction Creations. This is the reason that there is no such thing as inside or outside. It just "is". That is the reason I call what I am teaching you "what is".

According to a physics book that I read, Einstein believed that there is not an exit in what scientists call our galaxy, but there are many. He thought that if you flew a craft out as far as you could, you would be contained and could not fly any further. If you understand the principles, concepts and keys that I am teaching, you will see that it is not accurate. It's just that you are too small to see it visually, therefore you are calling it a galaxy instead of what it is. It is the make-up of a living thing or many in that case. You cannot really tell because when things are exponentially larger than you, and it is in slow motion from your perspective. If it is moving, you cannot really tell. It is seamless like our female conjunction parent's movements. You cannot really feel or tell that she is spinning and circling.

In the scheme of things, you may think that you are very small in such a grand scale according to the Law of Infinite Occurrences. You might feel like a grain of sand. However, every grain of sand is accounted for. Every grain of sand is important. It takes all of them to help form our spherical living thing. It takes each grain to help provide nutrients to different plants, trees, and blades of grass. There are different types for different things. It takes each individual grain to make the nest or house of a community of ants and other species. You are unique and needed for specific individuals with that special something that you came here with. You may not feel like you have it, but you do. You are created in a unique way for a reason. Always believe in yourself and remember that you do make a difference.

Learning that you were created by your fiery conjunction parent and non-fiery conjunction parents answers the question: Where did I come from?

This contributes a vivid image of where you come from. Not only that, but I informed you of how you were created by telling you how you were reproduced conjunctively. I also revealed your conjunctive parent's offspring. Since this is true, you know of entities that you are related to as a conjunctive sibling. All of this illustrates what you are. I have also shown you different likenesses that you have of your fiery conjunctive and female conjunctive parents. As a result, this gives you a sense and helps you to formulate who you are. In this case, this should make you understand yourself better. I allowed you to explore a world of opposites. Therefore, I have revealed how things work paradoxically and how this type of world that you live in was created. I have given you examples that go to infinity. Therefore, I have allowed you to see beyond the scope of any genealogist. I even gave you a bonus which is: an inside look at what scientists call our galaxy. This should give you a different outlook on life. Welcome to the beginning of your new journey of "what is" or how it really is.

Laws Discovered by Anthony

1. Law of Fast and Slow Motion
2. Law of the Span of Life
3. Law of Appearance According to Size
4. Law of Conjunction Creations Survival
5. Law of Infinite Occurrences
6. Law of Creation with Conjunction Creations

Helpful Resources

I create many things. If you would like to see a different type of innovative work of mine, please visit: https://www.AnthonyEntertainment.net. Of course, I would love for you the check out more books at: https://www.WhereDidIComeFromBook.com. I would also advise you to read *What Gave Me Life*? Which is the 2nd book in this series by visiting that website.

Subscribe to receive newsletters, bonuses, sales, offers, deals, new books, events, calendar dates, and more by clicking this link below:

https://www.wheredidicomefrombook.com/contact/